F**K
FAST FASHION

F**K
FAST FASHION

101

**ways to change
how you shop and help
save the planet**

TRAPEZE

First published in Great Britain in 2020 by Trapeze
An imprint of Orion Publishing Group Ltd
Carmelite House, 50 Victoria Embankment, London, EC4Y 0DZ

An Hachette UK Company

1 3 5 7 9 10 8 6 4 2

Text © Orion Publishing Group Ltd 2020
Illustrations by Emanuel Santos

A CIP catalogue record for this book is
available from the British Library.

Hardback ISBN: 9781409197980
Ebook ISBN: 9781409197997

Printed and bound in Great Britain by Clays Ltd, Elcograf, S.p.A

MIX
Paper from
responsible sources
FSC® C104740

www.orionbooks.co.uk

CONTENTS

INTRODUCTION

So, what is fast fashion?

Glad you asked! Fast fashion is the term used to describe any clothes that move quickly from the runway to the high street in order to meet new trends. As a result, the garments are speedily produced and sold at a lower price.

This revolution in shopping began in the 1990s, when retailers came up with the innovative idea of bringing the runway to everyday people. In fact, the term was coined by the *New York Times* to describe one, now global, chain's mission to go from designing a garment to having it available on the racks in only fifteen days.[1] This means that now, if there are puff sleeves on the catwalk at the latest fashion show, you'll see them on your high street the next week. Impressive, huh?

But let's take a step back. Fast fashion might seem like a dream come true but it comes at a cost. So, to really understand what fast fashion means, we need to take a look at the fashion industry *before* retailers hit fast forward on the whole process.

Once upon a time, clothes took a really, really long time to produce. You had to find the materials to make them, prepare them and then hand-sew or weave them into clothes. Or, if you weren't too nifty with a needle and thread, and could afford to, you paid your local tailor or seamstress to do it for you. Either way, it was seriously time-consuming.

But bring on the 1800s and Britain had its Industrial Revolution, which changed the way we produced garments for ever. How? By the invention of the *drumroll, please* power-driven sewing machine.[2]

This was game-changing. It meant that clothes and shoes could be produced on a large scale and, instead of waiting for tradesmen to create a garment for you, clothes could be produced ahead of time, ready-made.

What happened next?

Of course, to make clothing on a mass scale, you needed people to actually work the sewing machines. This led to the creation of sweatshops and, sadly, even from the beginning, most of these workplaces didn't offer the safe environment or conditions that are still being campaigned for today.

Towards the end of the twentieth century, then, clothes were being produced more easily than ever before but there was still a big difference between high-end and high street. Enter the 1990s and the dawn of fast fashion. New retailers were founded to produce affordable clothing that reflected high-end designs and the way we shop for clothes was transformed again.

Fast fashion: the problem

So here's where the problems began. Shopping used to be considered an event – you would save your hard-earned money and buy a very specific piece of clothing that you'd been coveting for ages. However, with fast fashion, shopping became a form of entertainment where consumers felt like they were wearing the same designs as the models at the fashion shows. This led to a throwaway culture because it was easier to buy new clothes and everyone wanted to keep up with the latest trend. Here are some headline stats:

👎 Across the world, we consume over **80 billion** new pieces of clothing a year (this is four times the amount we were consuming twenty years ago).[3]

👎 However, stats show that in the UK, **£10 billion worth of clothes are bought and not worn**[4] and in the US, the average American **throws away around 36kg of clothes a year**.[5]

👎 This means **1 rubbish truck full of clothes** (2625kg)

is thrown away **every second**, which is enough to fill the Empire State Building 1.5 times every day or Sydney Harbour every year.[6]

But the rise of fast fashion doesn't just create waste – it is founded upon poor working conditions for those making the garments and BIG issues for the environment too.

🌍 In 2015, the industry created **92 million tons of wastewater** (that's 20 per cent of global wastewater), leading to pollution of our water sources and soil.

🌍 The fashion industry makes **63 per cent of its clothes from petrochemicals**, which include plastic polymers. Common materials are nylon, rayon, viscose and polyester.[7]

🌍 What's more, every time a synthetic garment is washed, around **1900 microfibres** are released into the water. These are then ingested by small fish, who are eaten by big fish, introducing plastic into the food chain.[8]

🌍 The industry also has an annual carbon footprint of **1.2 billion tonnes of CO_2** – larger than the footprint of international flights and shipping combined. And when you take into account the impact of washing our clothes and throwing them away, the carbon footprint grows to **3.3 billion tonnes a year** (that's 8 per cent of global greenhouse gas emissions).[9]

What's to be done?

Lots! It's a pretty bleak picture but the brilliant thing is that, across the world, we are finally starting to pay attention to the impacts of fast fashion.

👍 At the 2019 G7 summit in France, the **Fashion Pact** was signed, committing thirty-two of the biggest names in the industry to reducing global warming, helping improve biodiversity and cutting down on plastic.[10]

👍 Launched in the UK, the **European Clothing Action Plan (ECAP)** aims to cut the environmental impact of clothes production. So far, it has been signed by Denmark, Finland, Germany, Italy, the Netherlands, Norway, Poland, Romania, Spain and Sweden.

👍 In Hong Kong, there's a growth in sustainable initiatives including the **Fashion Clinic**, which helps people mend their clothes by offering services and workshops.

👍 And in the US, the **#WearNext campaign** in New York is encouraging people to donate their old clothes to stop so many reaching landfill.

But it's not just governments, big companies and charities taking notice – the fight against fast fashion begins with individuals. Like you! And the changes you make don't have to be big ones either. By making one small change to how you shop, buy, wear and dispose of your clothes, you'll be making a massive difference. So, let's crack on, shall we?

SAVVY

SHOPPING

#1 It's vintage, baby!

Shopping vintage is one of the best (and most stylish) ways to reduce your wardrobe's carbon footprint and it's becoming increasingly popular – 64 per cent of women in 2019 agreed they were willing to buy pre-owned pieces, compared to 45 per cent in 2016.[11] Vintage clothes still have an environmental impact (most need some TLC, some will get wasted and the clothes still need to be transported from place to place), but this is minimal in comparison to the production of new garments. If you're not sure where to check out first, have a look online at ASOS Marketplace, Depop, The Stellar Boutique and Beyond Retro. It's time to embrace the shift to thrift!

#2 Charity-shop gems

There's nothing more satisfying than finding a stylish piece in a charity shop – not only do you help find unwanted clothes a new home, but your money is going to a good cause. Bravo! Oxfam, Emmaus and the Red Cross have stores across the world, but if you don't have a charity shop nearby look for local charities or organisations that are hosting a car-boot or second-hand sale. This means you're helping save our earth as well as getting involved with your community *ceremoniously places medal for 'best neighbour' over your head*.

#3 Local > Big Brand

There are so many local heroes who are using fabrics that are both sustainable and ethically produced. Next time you're hitting the shops, instead of making a beeline for the big brands, pop into your independent and see if they have what you need instead.

#4 Raiders of the lost outfit

Depending on how cheeky you feel, you might need to get permission for this one (or not). Some of the best clothing finds are closer to home than you might think. Like, really close. Head upstairs and have a snoop in your parents' wardrobe – chances are there's something from a few decades ago that's coming back into fashion (hello, shoulder pads *wave*). Even better, check out your grandparents' wardrobe. But do exercise caution – some things should be left in the past. (Looking at you, sweater vests . . .)

#5 Get inspired

Shopping retro is a great way to get your creative juices flowing, but if you're feeling a bit daunted by the prospect, check out some old photos and magazines to get some inspo. There are also lots of amazing Instagram accounts showcasing just how dreamy second-hand clothes can be. A few of our favourites are @pandorasykes, @sara_waiste, @op_shop_to_runway, @morethanyouraverage and @fixated_f.

#6 Do your research

Before you hit the shops, do a bit of digging into your favourite brands to see how sustainable their practices are. Three of the biggest things to look for info on are their manufacturing process, how they package their products (i.e. if they use eco-friendly materials) and their delivery options. If they don't provide any information on their supply chain, ethics or sustainability credentials, it's not a great sign. But do get in touch and ask them more – you'll feel empowered by taking matters into your own hands and if they're not already operating sustainably, it'll add pressure for them to reconsider their practices (for more on lobbying, see page 35).

#7 Pass on unwanted gifts

Reindeer socks? Again? Thanks . . . We've all been gifted those items of clothing we're just *never* going to wear (there's only so much Christmas-themed footwear a person can handle) but all is not lost. Have a think about who in your life might appreciate them more and pass those gifts on.*

Lots of homeless shelters look for warm clothes during the winter season, including socks, hats, gloves and coats, so do keep this in mind.

* Just remember the #1 rule – don't give them on to the person who gave them to you!

#8 Rent it!

One of the best ways to avoid buying items you're only going to wear once is by renting them. Perfect for weddings, proms, balls and more, companies like Rent the Runway, Mud Jeans, Girl Meets Dress and HURR are helping us take care of the environment while also looking our oh-so-stylish best.

If you really enjoy renting, there's no reason you can't do it for everyday outfits too. It also means you don't have to commit to keeping items for ever – perfect for those who like to follow the latest trends. Our top tip, though: avoid red wine while wearing . . .

#9 Complement your skin tone

When buying major wardrobe items (think coats, tops and bottoms), choose one or two colours that suit your skin tone and then purchase garments in these colours that can be worn interchangeably. The List has a great guide to helping you identify the colours that will suit you best on their website[12] and once you've got these colours sorted, you can find a few other pieces to spice up your base outfit and keep your wardrobe looking snazzy.

#10 Challenge. Accepted.

It can be hard making a change to your routine but, like lots of things in life, it's MUCH easier when you're doing it with others. Oxfam runs Second Hand September where you're encouraged not to buy anything new for thirty days, but if that feels too easy, there's also Fashion For Good's three-month challenge over summer.[13] Go on, create that new WhatsApp group now – we're aaall in this togetheeerr . . .

#11 Tote-ally worth it

You already know this one, but it's so simple, it's easy to forget! Never go shopping without a handy tote bag. Not only are these made from a more sustainable material (aka not plastic) and more durable, but they also look super snazzy. You'll find some beautifully designed ones at bookshops, museums and galleries as well as fashion retailers.

#12 It'll make you 'appy

Our phones can be one of the handiest tools for fighting fast fashion. A quick search online will give you lots of options for apps that will help you make sustainable shopping choices, but a great free one to download is Good On You. It rates over 2000 brands on an ethical scale of 1–5, using data from organisations such as Greenpeace and Carbon Trust, by considering the retailer's product choices and workers' conditions. If your favourite brand turns out to be less ethical than you hoped, the app gives

you recommendations for greener options and, what's more, it gives you the option of contacting the retailer to let them know they're not meeting your sustainability standards. Hear, hear!

#13 Avoid next day

It's very (very) tempting to click one-day delivery or same-day shipping when you have the option to, but this is not eco-friendly. When retailers promise to get their goods to you quicker, they often need to put on a special delivery (either by van or plane) which will likely have less cargo on board because it's sent separately from the usual deliveries. If you don't *need* it next day – go standard.

#14 Support other activists

By making a change (even if it's a small one) you – yes, you! – are an activist. You're helping to make our society more responsible and save our world. That's a pretty big undertaking and it can feel overwhelming but don't panic! It's cool. There are loads of people who are doing the same as you and fighting the good fight. If you ever feel like it's too much, or you start to forget why you're even doing this in the first place, check in with your favourite campaigners online or, if you don't know any yet, here are some to follow: @gretathunberg, @annehidalgo, @the_press_tour, @jen.brownlie, @treesnpeace, @sustainably_vegan, @mrspress, @liviavanheerde and @tollydollyposh. Have a browse, drop them a like or comment or even slide into their DMs, but most of all, remember: you're not alone.

"WE NEED YOU"

#15 Look at the label

Often there's lots of information included on the care labels of clothes, so when you're next having a browse, start by reading the label to see what fabrics the garment is made of and if it's recycled or organic. This is also a great habit to get into when it comes to taking better care of your clothes (see pages 69–90) and lots of brands include a note on their practices too, so get reading!

#16 It's time to unsubscribe

4551 emails?! From who? Liberate your inbox from those hundreds of fast-fashion retailer newsletters and free yourself from the relentless (and sometimes quite aggressive!) advertising that promotes unsustainable fashion. By hitting that unsubscribe button, you can cultivate your inbox so it's full of retailers who align with your values and make your inbox a happy place to browse. Click, click. Done!

#17 Eco-exercising

If you exercise a lot, you might find it tricky to avoid synthetic materials. Lycra, for example, is typically made from polystyrene, which is made from oil and releases microfibres into the water when washed. There are, however, an increasing number of sportswear brands that are making efforts to use recycled versions of polyester, nylon and Lycra. Check out Alternative Apparel, Miakoda, Girlfriend Collective and Nice to Meet Me for some sustainable alternatives next time you're in the market for some sporty gear.

#18 Shopping online – the good

More and more of us are choosing to shop online and recent figures show that in the UK online sales make up almost 18 per cent of all sales (in comparison to just 3.4 per cent in 2007).[14] It might seem counter-intuitive, but shopping online and getting your clothes delivered can actually be *better* for the environment than visiting stores in person. In 2010 a study found that you would have to buy twenty-four items in a shop to make the drive there and back as efficient as ordering just one piece online![15] This is because the delivery van acts like a bus for your clothes and some companies now even offer a 'green' delivery slot to ensure the routes are as efficient as possible.

#19 Shopping online – the bad

But online shopping isn't all sunshine and rainbows (sadly). With the internet comes so much choice that we're often buying waaaay too much – more than we need and definitely more than we'll ever keep (hands up if you've ordered three sizes of the same piece before *slowly raises hand*). It's estimated that in the US, 40 per cent of all online clothing purchases are returned[16] and in the UK, returns cost retailers £60 billion a year.[17] This means the carbon footprint for those garments is doubled for little reason at all.

Next time you're online shopping, only order exactly what you need and make sure to use any tools online retailers provide (like suggested sizes) to try to reduce the chance that you'll need to send something back.

#20 Shopping online
– the ugly

Unfortunately, the hidden story of returns gets worse . . . Often brands can't resell the items because the garment comes back damaged, unsellable (this can be as simple as a silk blouse that's too creased) or because it would take too long for them to process the item to then sell on. These clothes then end up in landfill, adding to our environmental impact. Added to this, each item will normally come packaged in its own plastic bag that can't be reused.

When you're trying on clothes, be careful not to damage anything so if you need to return it, there's a higher chance it can be sold on.

#21 Pop to the shops

Online shopping isn't always better for the environment, though. If you live or work within walking distance of the shops, or can travel there by public transport, visiting the store in person is a much greener option. There's also the added benefit of being able to try the clothes on while you're there, making it less likely you'll need to return anything.

#22 Thoughtful shopping

Wherever you shop, it's important to stay mindful while browsing. Some retailers have a 'one-click' option which can lead to impulsive shopping (and therefore returns). Before you click 'buy', check your basket, make a cuppa and come back – do you definitely want all of those items? If so, go for it! But if not, wahoo – you've just saved yourself the hassle of returning something and won planet-saving points.

#23 Small but mighty

Most of us will shop at the same three retailers because hey, who likes change? But the time has come to be more adventurous and branch out from the high street. There are lots of online independents that are committed to making a change – here are some brilliant ones to get you started:

Newbie (children's)
Birdsong
Palava
Thought
Know the Origin
Komodo
People Tree
Boyish Jeans (jeans)
Project Pico (lingerie)
Luva Huva (lingerie)
Matt and Nat (accessories)

#24 Smart accessories

Ever looked at your outfit in the mirror and thought it's missing . . . something? Chances are you need some kick-ass accessories to bring your outfit together. Invest in quality pieces (think jewellery, hats, scarves and gloves) that will go with lots of different outfits. This will help keep your capsule wardrobe looking fresh while avoiding buying unnecessary additional clothes.

#25 Old favourites

Alternatively, if you've got a necklace that you haven't worn in months, think about what clothes in your wardrobe could go with that piece. It'll help you think creatively about what you already have and ensure no hidden gems are left forgotten in your jewellery box.

#26 Swapping sites

There are loads of websites now that allow you to connect with others nearby who are getting rid of clothing either for free or for something of similar value in return. This is a great alternative to recycling and especially useful if you're on the lookout for something in particular. Do a search for 'clothes swapping' online to find a platform that operates in your area.

#27 Speak up!

Thankfully, protecting our environment is now a worldwide conversation and there are lots of ways you can lend your voice to the cause. If you're ready to pick up your placard, check out Extinction Rebellion, who organise protests throughout the year. Alternatively, if you live in a remote area or are not a fan of marches, get involved on Twitter and keep an eye out for trending hashtags when events are going on.

Here are some hashtags to find out what's happening right now:

#environment
#nature
#climatechange
#savetheplanet
#ecomonday
#greentweets
#activism
#sustainable

#28 Keep them in the loop

Let your friends and family know about your fight against fast fashion. We don't mean this in a humblebrag sort of way (trust us, it won't go down well with your pals), but instead tell them about your passion for anti-fast fashion so they don't buy you any gifts you don't want. If it's your birthday or a special occasion and they would like to buy you clothes, send them a list of the brands that you support, so they can get you a present both you and the world will be grateful for.

#29 Savvy Secret Santa

Next time you and your mates organise a Secret Santa, make it a rule that you can only buy second-hand gifts for one another. There are so many fantastic pre-loved pieces to be found (check out tips on pages 9–10 and 13) and you'll be surprised what you can buy if you're on a budget too.

TO BUY

OR NOT

TO BUY?

#30 Think 30

Here it is. THE golden rule when you're at the shops and choosing what to buy – will you wear this item at least thirty times? Yes? Grand, pop it in the basket. If not, put it back and walk on by. As you know, we've got a big problem with waste (in the US alone, over 11 million tons of textile waste are produced every year[18]) so tackling the problem head on from when we first choose to buy something is key. Try to avoid statement pieces that you think you'll only wear once and if you've got a special occasion coming up, check out our tips on pages 16 and 46 on how to sustainably dress to impress!

30?

#31 Create a capsule wardrobe

It's much easier than you think! When you're buying new items, try to choose pieces that are timeless in shape and colour and will match with lots of other pieces you have (sorry, lime-green leggings straight from the runway) – that way, you can create lots of looks from fewer items.

Your capsule wardrobe can be made of ten pieces or up to fifty – it's all about finding the right balance for you. Maybe you can create a capsule wardrobe[19] just from the clothes you already have? But if not, when you're out buying, approach each piece like you would a haircut – does it work on its own and can you dress it up and down? If yes, chances are it deserves a coveted spot in your capsule.

#32 Sunny with a hint of snow

Sadly, for most of us it's not a balmy 30°C all year round, so when it comes to buying new pieces, consider what it would be like wearing this piece during the rest of the year. Of course, there are some items that are only suitable for certain seasons (good luck trying to make a bikini work during the colder months) but if it's a top, skirt, trousers, jeans or coat, think: can you layer this piece up or down to accommodate different weathers?

#33 One in, one out

Yep, you heard right. This is one of the most brutal rules in the book but it's a foolproof way to keep your shopping habits in check and your wardrobe a manageable size. If you're buying something new, think about what you are willing to recycle to make room for it. This top tip will ensure that when you're buying something, you're making a considered decision and checking whether you really, *really* need that item.* It's good for the soul too – check out page 45 for the mental-health benefits of a smaller wardrobe.

* It also makes it much easier if you ever decide to move to a new house – nobody has time to pack up a million and one pairs of jeans.

#34 Time to get naked?

As with most things in life (except perhaps chocolate and videos of puppies), less is more, and clothing is no different. It takes more than 5000 gallons of water to manufacture just a T-shirt and a pair of jeans,[20] so not only is a smaller wardrobe better for your mind, but it's also better for our planet. Having less can be really freeing (less laundry, folding, picking up, finding space in the chest of drawers, etc.) so as you browse the shops, keep a minimalist state of mind and, who knows, maybe you'll find it so liberating you'll end up living in a naturist community . . .

#35 Give yourself time

Even with the best intentions, it's hard to make eco-friendly choices when you're in a rush (especially if you're trying to buy an outfit for your best mate's wedding the day before the big day). To combat this, try to start thinking about any new purchases you might need for future events sooner rather than later.

#36 The spending revolution

As a society, we're encouraged to spend loads of money on clothes we'll only wear a few times (think wedding outfits and party dresses), but next to nothing on our everyday items (like vests, knickers and tops) – bizarre. In terms of cost per wear, that logic just doesn't work, my friend, and it's a big part of how we've ended up in this sticky, wasteful mess. It can be hard to put into practice, but when you're out shopping, really question your spending habits and try to fork out more on workwear and underwear, i.e. those pieces you'll wear most, and less on those you're only going to wear for special occasions. In the long term, you'll buy (and waste) less. *Vive la révolution de la mode!*

#37 Fabrics to avoid

Alas, not all materials are born equal and some are especially damaging for our planet. If you can, try to avoid these fabrics where possible unless they're needed for a functional purpose (like making your coat waterproof):

- Rayon
- Spandex
- Nylon
- Acrylic
- Modal
- Polyester

They've all got a high environmental impact since they're non-biodegradable and made from synthetic materials that shed microplastics when washed. (Check out page 50 for our hero fabrics.)

#38 Time to cotton on

It's not just artificial materials that can be harmful to our environment. Natural fibres like cotton also require pause for thought.

Cotton is the most profitable and widespread non-food crop in the world and the industry provides jobs for over 250 million people, but the way it's typically farmed is unsustainable. Its production requires huge amounts of agrochemicals (like fertilisers and pesticides), over 20,000 litres of water and often results in the destruction of habitats and degradation of the soil.[21] This means ecosystems are being lost and people working on these farms are suffering from exposure to dangerous chemicals.

Of course, it's hard to avoid cotton completely but being aware of how it's created and other more sustainable and ethical options is a step in the right direction!

#39 Hero fabrics

We know what to avoid, so here's what to look out for! All of the below are either sustainably sourced or considered to have a low impact on the environment. Some are easier to find and more affordable than others, but by choosing just one of these materials instead of another, you're making a difference. After all, every little helps.

- Wool (also see page 51)
- Silk
- Hemp
- Linen
- Tencel (a sustainable option made from wood fibres)
- Piñatex (an innovative leather substitute made from pineapple leaves!)

#40 Un-baa-lievable

Ewe've guessed it, wool is one of the most sustainable fabrics we can wear so it deserves a special mention. Not only is it renewable, biodegradable and recyclable but it's also durable, has natural flame-resistant qualities (so no added harmful chemicals are needed) and is water repellent. When you're browsing for new clothes, look for wool-based options (or, even better, organic wool) and if you're crafty you can even knit your own clothes. From hats and scarves to sweaters and dresses, you'll find lots of patterns online at sites like www.allfreeknitting.com and www.ravelry.com. Now we just need to find our knitting needles . . .

#41 Certification clues

Knowledge is power and you're already following this motto by reading this book! Another way you can empower yourself is by getting familiar with the different certifications the fashion industry uses to help you make more informed decisions while out shopping. Here are a handful to look out for:

- Global Organic Textile Standard: clothes made with a minimum of 70 per cent organic fibres.
- Forest Stewardship Council: clothes made from sustainably harvested trees.
- Fair Trade and Fair Wear Foundation: clothes created in conditions where factory workers are paid fairly and are safe.

#42 Natural dyes

Lots of traditional dyes contain toxic chemicals that need huge amounts of water to be processed. When washed, these chemicals are released from the fabric and pollute our water systems, making textile dyes the second largest polluter of clean water in the world.[22] To combat these devastating stats, look for natural and low-impact dyes such as those made from vegetables (like natural indigo and saffron) and insects or animals (like cochineal). Also opt for unbleached materials. Pure white fabrics, although looking as if they might be undyed, are often bleached with chlorine.

#43 Recycling champion

We all know the importance of recycling, but you win extra points if you can buy products made from recycled materials. You 1–0 Fast Fashion. There are loads of smaller companies making waves by using recycled materials in their garments, like Lucy & Yak whose fleeces are made from old plastic bottles. Some bigger companies like Patagonia, Stella McCartney, Weekday and Nike are starting to get involved too. Check out the websites of your favourite and local retailers to see if they've got any lines that are made from recycled materials.

#44 Wear it and wear it again

We've all been there – you're getting dressed in the morning and you want to wear that pair of trousers *again*, but what if someone remembers you wore them yesterday? Do you need to wear something different?! Chill. We need to shift our attitude. It's totally fine to wear something two days in a row and, let's face it, nobody's really paying that much attention to what you're wearing anyway (unless you're a celeb hitting the red carpet). Embrace wearing your outfits at least twice and, as a bonus, enjoy the time and energy you save in the morning not having to fuss over what you're going to wear that day.

#45 Money, money, money

Sadly, money doesn't grow on trees (and if it did, you'd keep that tree's location a secret, huh?) so before you buy, ask yourself this: is there anything you'd rather spend your hard-earned money on? It's very easy to get caught up online or in the shops and spend more than you planned to, but maybe something else would bring you more joy. Channel your inner Marie Kondo and question what would make you happier: this new turtleneck jumper or dinner somewhere special? Make your decision and own it.

#46 You gotta have patience

One of the allures of fast fashion is how quick and easy it is to buy clothes, but if we want to make a change, we've got to have some good ol'-fashioned patience. It might be that you need to save a bit more money to spend on a better-quality item or you need to do a bit more research to find a retailer that uses sustainable fabrics and sells a top you like. Embrace the power of slow and enjoy the warm glow of knowing you're doing your bit for our planet.

#47 If it seems too good to be true . . .

. . . it probably is. (No shocks there.) It's an inevitable fact that quality items are more costly to produce so this additional cost is passed on to the consumer through a higher price tag. If you've found some clothes that claim to be sustainable but are ridiculously cheap, it's worth doing a bit more research to see if everything checks out.

#48 It's sales o'clock

OK, hitting the sales miiiight be your worst nightmare but without some risk, there's no reward. In this case, you might lose your sanity BUT there is simply no better time to grab yourself a bargain. For the sales novices among us, typically spring and summer items will go on sale in June and July and autumn and winter clothing will go on sale in January. However, with the dawn of internet shopping, you can bag a bargain at any time of the year. Once you've identified your favourite brands, sign up to their newsletters so you're first to know when their products are reduced and then bask in the smugness of knowing you're getting a top-quality item for less.

#49 Be a quality detective – part 1

Buying better-quality items means that they'll last longer, you'll buy fewer items and therefore less clothing is wasted. Win. Win. Win. There are lots of things to look out for, but here are some to get you started:

1. Opt for metal zips over plastic – they're much sturdier and more durable.
2. Spare buttons? Great, it shows the manufacturer believes their product will last long enough for you to need extras.
3. Give the stitches a gentle tug – they should hold firm and not pull apart.
4. No loose threads. Make sure to have a thorough check for these troublesome fellows. Like movie franchises, they only get worse.

#50 Be a quality detective – part 2

Don your deerstalker hat and pick up your magnifying glass – here are more checks you can do when buying your clothes to make sure that you're getting quality items that will stand the test of time:

1. Does it fit properly? Often, cheaper clothes cost less because they use as little material as possible, so try to avoid anything with too-short sleeves or legs (unless that's what you're looking for) or tight shoulders.
2. Does the pattern match at the seams? This might not be something that bothers you aesthetically, but higher-quality clothing will make sure this detail is adhered to.
3. Are the seams finished properly? Make sure the garment lies flat and it doesn't pucker at the seams. Also turn it inside out to check there are no unfinished edges.
4. Check the thickness of the material by holding

it up to a light. Higher-quality fabrics are usually thicker.

#51 Selfie time

It can be tricky deciding whether or not you think you look just good or FABULOUS in a new piece of clothing. To help you decide, snap a quick selfie and see what you think. If you're feeling cute, it's a good sign that this is an investment worth making; if you're more 'meh', hand it back on your way out.

#52 Focus on your feelings

It's so easy to focus on how clothes look on you that often we forget to take a moment to think about how different clothes make us *feel*. When you're out shopping, try to be mindful of whether that jumper makes you feel happy and cosy or uncomfortable and itchy. If it doesn't make you feel good, the likelihood of you ditching it in a couple of weeks is high. Next time you're in the changing rooms, close your eyes, count to ten and check in with yourself before you buy.

#53 Don't give up!

By reading this book, you're taking part in helping to save our environment *highest of fives*. But combatting fast fashion is a big commitment and nobody is expecting you to completely overhaul your wardrobe overnight. Make one change at a time and if you fall off the bandwagon, show yourself some kindness. By doing something, you're already a superstar!

WEAR

AND

CARE

#54 Let's get organised

Here's the thing. We will nearly always do what is easiest. So, when it comes to picking out clothes, you're always more likely to choose the clothes that you can see *side-eye glance to that heap of clothing next to the bed*. It's time to get organised.

If you have a wardrobe, hang your clothes up (excluding any that should be folded, see page 90) and if you like, arrange by colour or length. If you have a chest of drawers, make sure your clothes are folded neatly and preferably so that you can see each garment.[23]

#55 How to care for: denim

The well-known jeans brand Levi Strauss states on their website that 3781 litres of water are used to create and care for just *one* pair of their jeans and that 33.4kg of CO_2 are created during their lifetime – that's the equivalent of 69 miles driven in the average US car.[24] With such a sizeable environmental impact, it's really important we take care of our jeans so that they last longer. Here are the three golden rules for dreamy denim:

1. Always wash inside out with cold water and similar colours.
2. Don't dry clean! Line-dry instead.
3. Avoid washing them too often – around every ten wears is ideal.

#56 Stain, stain go away

We've all had that heart-sinking moment: you're wearing a white top and that plate of spag bol is suddenly looking like a very bad idea . . . seconds later, it's bye-bye-blouse time. But wait – you packed your stain removal pen! These nifty little pens are perfect for reducing the chance of permanent damage when you stain your clothes while out and about. A handy investment that won't break the bank.

#57 Pattern protection

If you've got a top with a pattern ironed on, make sure to wash it inside out. This will help avoid fading and cracking, so your statement pieces stay looking sharp.

#58 Washing wisdom

Manufacturers know their garments best and through the labels they're equipping you with all the information you need to ensure your clothes stay looking brand new for as long as possible. The detergent company Ariel has a handy online guide if you're not sure about what the different symbols mean[25] and if you're prone to forgetting, print out the guide and stick it somewhere near your washing machine. Your clothes (and the planet) will thank you!

#59 Be prepared to repair

If a button's come loose, a stitch has come apart or even if you've got a hole in your well-worn trousers, think about repairing them before you head to the recycling bin. It's surprisingly easy and affordable to fix minor wear-and-tear problems and there are lots of videos on YouTube that can give you some guidance if you're not too savvy with a needle and thread. Alternatively, you can always pay a professional to fix it for you – either way is much cheaper than buying something new and it's better for the planet.

#60 Stomp!

Out of all your clothes, your shoes get a pretty hard time and are likely to be the first to show signs of wear, especially if you use the same pair day in, day out. Keep an eye on the heels and if they start getting close to the base, nip to your local shopping centre and get them reheeled. It's cost effective and saves the pain of wearing in a new pair of shoes . . .

#61 Keeping in shape

When in storage, help your shoes keep their shape and last longer by storing them with balls of newspaper inside. There's a reason why the shoe shops do it!

#62 Wash at 30°C

Washing at a lower temperature is often recommended for delicate clothing (like wool and silk) but it's also helpful for preserving the colour in dyed materials and, most of all, it uses less energy. It was once thought that washing at a lower temperature meant your clothes weren't cleaned as effectively, but with more advanced detergents now available, there's no need to worry (unless you have an especially tricky stain or are washing towels – then you might need to opt for a hotter wash).

A recent study found that if every household in the UK turned their washing down from 40°C to 30°C for a year, it would reduce CO_2 emissions by an amount equivalent to powering 1550 homes for a whole year.[26] Impressive, huh?

#63 Don't overfill

You're down to your last pair of pants. The pile of washing is like a mountain at the end of your bed. Do it all in one go? Nope. Think again. Overfilling your washing machine causes the clothes to rub against each other, meaning they'll fade faster and even catch on the drum and tear if super full. Try to keep on top of your washing by doing it regularly (yes, when you have at least three pairs of pants left clean).

#64 But also, don't underfill

Yep, washing machines are like Goldilocks and her porridge. You've got to get the load just right. Underfilling your washing machine equates to a big waste of water. The average home will wash 400 loads of washing every year, equating to 60,000 litres of water, so you want to make sure every drop of water is put to good use.[27] Aim for a full but not overflowing load and you'll be washing like an eco-warrior in no time.

#65 Vinegar vs stains

This versatile liquid doesn't just belong in the kitchen – it's a great stain remover on leather and suede. Happy days!

#66 Avoid dry cleaning

It might be convenient but dry cleaning involves harsh chemicals that have a negative impact on the environment and cause the material to fade quicker. Instead, hang up your clothes (including suits) in a steamy bathroom while you take a shower.

#67 Suede and leather

If there are two materials that don't do well in wet weather, it's got to be these two. Keep them safe in the rain by using a non-toxic waterproof spray and refreshing as instructed on the canister.

#68 How to care for: delicates

Crying over your warped lace knickers isn't a good look. So, to avoid a delicates disaster and ensure they last longer, follow these three easy steps:

1. Check the label.
2. If it says handwash, do it. It's a bit more of a faff, we know, but here's a beginner's guide if it's your first time: fill your sink or bath with cool water and add a gentle detergent. Separate your delicates into similar colours and add to the water, following the detergent's instructions. Lightly squeeze when done and dry on a rack.
3. If you're really not keen on hand washing, the next best option is a delicates bag. You can buy these online or at large supermarkets, but remember to use your washing machine's shortest and most gentle wash and spin cycles to ensure minimum damage.

#69 Say hello!

If you've found a brand that's doing a great job promoting sustainability or has made a recent step in the right direction, let them know you've noticed! Retailers not only take note of constructive feedback, but they also pay attention to positive comments, so make sure to tweet, email or phone the retailers you think are smashing it.

#70 Protect your colours

As a general rule, wash your dark clothes inside out – it helps the colour last longer and prevents them from colouring your other clothes.

As well as washing them inside out, help your jeans keep their colour by adding two tablespoons of salt in the machine when you first give them a wash.

#71 Three, it's the magic number

Try to wear your garments three times before throwing them into the laundry basket. Unless they're smelling, erm, potent (no judgement here) or are visibly dirty, it's fine to wear them again before washing. If you do find that your clothes aren't smelling too fresh by the end of the day, air-dry them overnight to help keep the air flowing.

#72 Baking-powder power

Detergents are unfortunately pretty harmful for the environment since they include phosphorus. When this chemical builds up, it leads to eutrophication (where an ecosystem becomes overly nutrient-dense), which encourages large amounts of algae to grow. Excessive algae causes oxygen depletion and, sadly, the deaths of fish and other living creatures.

OK, that's a pretty dire scenario. However, this household trick will cut your detergent use massively. Next time you put a wash on, use half a cup of detergent with half a cup of baking powder. Your clothes will still be cleaned effectively and come out smelling like roses. Dreamy.

#73 Sun strength

If it's sunny outside, pop your clothes on a line and dry them outdoors. This way, you avoid the energy used for powering your dryer and it also gives you the chance to get some fresh air and good ol' vitamin D. Just remember to give your line a wipe down before hanging up any clean clothes and give everything a little shake before bringing it back in.

#74 Lipstick comes last

Your 'Ruby Red #3' might look gorgeous on your lips, but not so much on your white dress. Avoid stains and marks on your clothes by getting dressed before you put your make-up on. Pucker up!

#75 Sensible storage

Some clothes simply don't do well on a hanger and this includes sweaters, heavy pieces and baby clothes. By storing these flat, you avoid any misshaping which means they'll last much longer and you won't be wondering why your favourite sweater has gone wonky.

REUSE
REVOLUTION

#76 Host a clothes swap

Have you always liked Jamal's top? Or Mo's jumper? Now's the time to seize your opportunity. Get all your pals around and host a clothes swap. The concept is simple – everyone brings their unwanted clothes and lays them out on a big table (or in a big pile) and 3...2...1...GO! Get stuck in, try on the pieces that catch your eye and go home with a bagful of goodies knowing that your old clothes have gone to loving new homes too.

#77 Riches to rags

If you're following the anti-fast-fashion lifestyle for long enough, it's likely some of your clothes will eventually tear or get a hole. But the end is not nigh! Torn clothes make perfect rags for cleaning the house. Simply cut into smaller pieces and put aside for your next spring clean.

#78 Let's get personal

Freshen up your old clothes by adding some personal touches. DMC Embroidery have loads of patterns online for jazzing up old tops or even hats with thread and if that's too tricky, cross-stitching is a simple and super effective alternative. Try to avoid adding plastic to your clothes (like standard sequins) and if you fancy a change of colour, use natural dyes (see page 53) to create a completely unique piece of clothing.

#79 Defeat draughts

This is a double whammy for reusing old clothing. An old pair of jeans is perfect for making the outside of the draught excluder. Simply cut off one of the legs and sew up the top end. Then stuff it with all your other old pieces of clothing and sew up the bottom.

#80 Bi-annual clear out

Every six months, set aside one day to reset your wardrobe. Go through everything (clothes, accessories, shoes, you name it) and put to one side anything you've accumulated but no longer want. If you're struggling to decide what to keep or not, think about whether you've worn it in the last year – if you haven't, perhaps it's time to let go and make room for something you will wear.

#81 Second lives

Lots of old clothes make perfect around-the-house outfits. For example, if you have some old dirty trainers, these make perfect shoes for gardening. Similarly, a stained T-shirt is ideal for when it comes to redecorating.

#82 Keep note

Once you've decided on the clothes you're going to give away (or sell), make a mental note of what they are. If you're always getting rid of sparkly shorts, think twice before you buy another pair – this way, you'll be preventing bad habits instead of repeating them.

#83 Make a quilt

We've all got those T-shirts that have sentimental value but we know we'll never wear again. Here's a lovely way to keep them in your life while also keeping you snug in the cold weather:

1. Get all your T-shirts in a pile. Sixteen makes a good-sized quilt.
2. Cut out a 30 x 30cm square from the middle of each shirt.
3. To keep your quilt strong, apply fusible interfacing to the back of each square, following the pack's instructions.
4. Lay the squares out in the shape and order you want – maybe you can arrange by colour in a chequered pattern or oldest to newest. Get creative!
5. Start with two squares and pin the joining edges together so the front sides face each other.

6. Using either a sewing machine or by hand, sew the squares together, leaving a 13mm gap at the top.

7. Repeat this process for the rest of the row, and then repeat for the other rows. Iron each row so the seams lay flat.

8. Cut a piece of backing that is 5cm bigger than your whole quilt on all sides.

9. Pin the T-shirt layer to the backing and stitch together. Ta-da!

#84 Napkins and runners

If making a quilt isn't your jam (or perhaps feels like a bit too much work), you can make napkins or table runners instead. To make a runner, simply follow the instructions on the previous page but for only one row (be sure to measure your table beforehand so the runner is long enough to hang over each side). To make napkins, keep the squares separate.

These make perfect presents too, especially for newly engaged couples. Embroidering their initials onto the corner of the napkins makes a beautiful personalised gift and you can be sure that no one else will have bought them the same thing!

#85 Snagged threads

Like a spot on your chin or a scab on your knee, it is SO hard not to pick at a snagged thread. Instead, if it's knitted, take a look to see if you can pull the loop back through on the other side or sew a knot over the loose thread and cut away the excess to secure it down.

#86 Sell, sell, sell

You don't have to be the slickest salesperson to make money from selling clothes nowadays. It's so easy to sell your clothes online – you don't even have to talk to anyone! There are plenty of websites that will help you sell your old clothes – check out Depop, eBay and Vinted, and watch the money roll in.

If you prefer a more old-school approach, host a garage sale or set up at a car-boot sale and get haggling.

#87 Get in touch with the manufacturer

A growing number of retailers are accepting their own clothes back once they've come to the end of their lives. Some of the brands leading the way are Patagonia, The North Face, H&M, & Other Stories and Levi's. Some retailers will even give you money back for returning old clothes (if you needed even more of an incentive!).

#88 BOO!

Halloween is one of the worst times for fast fashion, with so many one-time-wear outfits being sold and most being made from plastic and other synthetic materials. Next year, get creative with what you have or, if you don't have any old clothes that you're happy to convert into a costume, check out your local charity shop.

#89 Avoid the bin

31st October is pretty scary, but y'know what's even scarier? What happens when you throw clothes into your bin. Textiles are thought to be the most harmful material after aluminium that end up in household bins. Not only do they release lots of toxins but they often take hundreds of years to decompose (if they decompose at all).[28] Always try to find another home for your clothes instead of the bin – once you've finished this book, you'll have plenty of ideas!

#90 Material memories

If you've got an old band T-shirt, football T-shirt or something signed, a great way to keep it in your life when you can't wear it any longer is by framing it. It's super simple – buy a frame (try to avoid plastic ones if possible), take the mount out and draw around its outline onto the shirt. Cut out the square/rectangle and tape the material to the back of the mount. Pop it back in the frame and, final step, hang it up on your wall.

#91 Get snipping

Old trousers are only a couple of swift cuts away from being a savvy pair of shorts – perfect if you've scuffed the bottoms or are in desperate need of something cooler for summer. Old knitted dresses can also be made into lovely cardigans and long-sleeve shirts can be given short sleeves. Just don't get *too* scissor happy and make sure you haven't accidentally picked up your sibling's top!

If you don't feel too confident about your sewing skills, investigate a simple sewing workshop online or somewhere local. It'll give you the confidence to get creative with your old pieces of clothing.

#92 Compost heap

If you've got clothes that are only made from natural (biodegradable) materials and dyes, you can have a go at composting them. These include types of cotton, silk, wool, cashmere, hemp, bamboo and linen. 1 Million Women have a great guide on how to do this but, in short, you'll want to cut the fabric into small pieces and make sure they're spread out evenly on your pile. Make sure you remove anything that's non-biodegradable (e.g. buttons, zips and labels) and use hot compost to make the magic happen faster.[29]

#93 Puffer-jacket magic

If your favourite puffer jacket has lost its puff, next time you're having a shower, hang it up in the steamy bathroom and it'll be full of life in no time.

#94 Upcycle old clothes

As well as turning your clothes into completely new garments (see page 109), you can also completely reinvent old pieces with some simple little tricks. Here are our favourites:

- Replace the old buttons with new (or vintage) ones.
- With tracksuit bottoms, replace the string in the waist with a contrasting colour material.
- Bored of plain white plimsolls? Get drawing (or, if you're not arty, ask a friend who is!).
- Add lace to the bottom of your top to make it extra fancy.

#95 Sleepy time

Turn your old T-shirt into a dreamy pillowcase by following these easy-peasy steps (it works really well with T-shirts that have a central design):

1. Turn the T-shirt inside out and stitch a rectangle slightly larger than your pillow, going up one side, across (under the neckline) and back down the other side. Make sure to backstitch at the bottom edge of the shirt and at the corners at the top to make sure they're super strong.
2. Trim away the excess material with some scissors. (See page 94 or 96 for a great use of the leftovers.)
3. Turn the pillowcase the right way around and pop in your pillow. Sweet dreams!

#96 A specs-tacular idea

Next time you need a new pair of glasses, consider getting just new lenses instead of frames. Chances are you have a drawer full of old pairs and now is the time to dig them out and show them off again. Most opticians will create new lenses for your frames (no matter where you bought them from) or have a recycling point if you don't fancy them any more.

#97 Ready, steady, cook!

Old shirts can be made into the perfect aprons. Here's how:

1. Cut along one of the side seams from the bottom of the shirt, then along the back of the armhole seam to the top. Cut across towards the collar (but not through!)

2. Repeat on the other side of the shirt. Then cut along the back of the collar so the back is removed.

3. Then remove the arms from the front by cutting at a diagonal from the armpit to the collar edge that's closest on both sides (try to make sure they match).

4. Finish the edges by folding under about 1cm of material and stitching along the cut edges.

5. Using the back of the shirt or the sleeves, cut two long strips and finish the edges as per the previous step.

You now have a working apron! If you want to go the extra mile, you can even create a handy pocket.

#98 Sharing is caring

If you've got a special event coming up and know that Prisha's top would be PERFECT for it, ask if you can borrow it for the night. It's like renting but without the hassle and you can always return the favour by lending something back. But again, it's probably best to avoid red wine . . .

#99 One for the animal lovers

Pets will always be very grateful for your old towels, bedding and blankets. If you aren't lucky enough to have a furry friend, check in with your local vet or shelter to see if they're accepting donations.

#100 'Armless shopping bags

Vests can be made into super shopping bags simply by
sewing up the bottom and using the shoulder straps
as handles. If you have a string vest, even better!

#101 Spread the word

Now that you've learned all these tips and tricks, go forth and share the message. It's only by working together that we can create a lasting change and by telling just one more person, you're making a BIG change. Of course, you don't have to shout about it from the rooftops (unless you want to!) but next time you're on the phone to your mum or catching up with the gals, share one of these handy tips. Perhaps they'll go on to tell one of their friends. The movement is growing – thank you for being part of it.

REFERENCES

1 Solene Rauturier, 'What is Fast Fashion?' www.goodonyou.eco (accessed 05/11/19).

2 There's a long and fascinating history of how the sewing machine was invented (lots of people tried to get it right!). You can read more about it in this article by Mary Bellis, 'History of the Sewing Machine' www.thoughco.com (accessed 05/11/19).

3 The University of Queensland, 'Fast fashion quick to cause environmental havoc' www.sustainability.uq.edu.au (accessed 05/11/19).

4 Sandra Halliday, 'Britons have £10bn worth of unworn clothes in wardrobes says survey' www.ukfashionnetwork.com (accessed 05/11/19).

5 Rick LeBlanc, 'Textile and Garment Recycling Facts and Figures' www.thebalancesmb.com (accessed 05/11/19).

6 Elizabeth Reichart and Deborah Drew, 'By the Numbers: The Economic, Social and Environmental Impacts of "Fast Fashion"' www.wri.org (accessed 05/11/19).

7 Kianna, 'Fast Fashion Facts: What you need to know' www.7billionfor7seas.com (accessed 05/11/19)

8 Sustain your Style, 'The fashion industry is the second largest polluter in the world' www.sustainyourstyle.org (accessed 05/11/19).

9 Jasmine Chinasamy, '"A monstrous disposable industry": Fast facts about fast fashion' www.unearthed.greenpeace.org (accessed 05/11/19).

10 You can read more about the Fashion Pact in this article by Fashion Revolution, 'The G7 Fashion Pact: What it is and what it's missing' www.fashionrevolution.org (accessed 05/11/19).

11 Amy Louise Bailey, 'From Fendi Baguettes to Chanel Boy Bags, These Are the Best Shops in the World for Designer Vintage' www.vogue.co.uk (accessed 04/11/19).

12 Check out The List's guide here: www.thelist.com/20621/best-clothing-colors-skin-tone/ (accessed 04/11/19).

13 Find out more about both of these challenges here: Second Hand September at www.oxfamapps.org/secondhandseptember (accessed 04/11/19) and @slowfashionseason on Instagram.

14 Rebecca Milligan, 'Online vs high-street shopping – what's more energy efficient?' www.energysavingtrust.org.uk (accessed 29/10/19).

15 Julia B. Edwards, Alan C. McKinnon and Sharon L. Cullinane, 'Comparative analysis of the carbon footprints of conventional and online retailing', International Journal of Physical Distribution& Logistics Management, Vol. 40 No. 1/2, pp. 103–123. DOI: 10.1108/09600031011018055 (accessed 29/10/19).

16 Elizabeth Segran, 'Your online shopping has a startling hidden cost' www.fastcompany.com (accessed 29/10/19).

17 Aliya Ram, 'UK retailers count the cost of returns' www.ft.com (accessed 30/10/19).

18 Planet Aid, '8 Little Facts About Our Clothing Habits' www.planetaid.org (accessed 04/11/19).

19 The term 'capsule wardrobe' was coined by British boutique owner Susie Faux after she created the notion to help women edit their wardrobes to include only necessary, high-quality pieces that could be worn with others. Read more in this article by Sarah Young, 'Capsule Wardrobe: What it is and why do I need one?' www.independent.co.uk (accessed 28/10/19).

20 Glynis Sweeny, 'It's the Second Dirtiest Thing in the World – And You're Wearing It', www.alternet.org (accessed 04/11/19).

21 Read more about the cotton industry here: www.worldwildlife.org (accessed 04/11/19).

22 The largest pollutant is agriculture. Read more in this article by Patsy Perry, 'The Environmental Costs of Fast Fashion' www.independent.co.uk (accessed 29/10/19).

23 Marie Kondo has an incredibly helpful video on how to fold your clothes so you can see each one here: 'How to Fold Your Clothes With Marie Kondo', Real Simple www.youtube.com (accessed 04/11/19).

24 Levi Strauss, 'The Life Cycle of a Jean' available online here: www.levistrauss.com/wp-content/uploads/2015/03/Full-LCA-Results-Deck-FINAL.pdf (accessed 04/11/19).

25 For a simple guide to understanding laundry symbols, check out Ariel's guide, 'Washing Symbols Explained' www.ariel.co.uk (accessed 28/10/19).

26 Jack Peat, 'Households are using twice as much energy as needed for laundry, reveals research' www.independent.co.uk (accessed 29/10/19).

27 Sustain Your Style, 'How can we reduce our Fashion Environmental Impact?' www.sustainyourstyle.org (accessed 29/10/19).

28 Megan Sutton, 'Had enough of fast fashion? Here's how to make your wardrobe more sustainable' www.goodhousekeeping.com (accessed 28/10/19).

29 For easy-to-follow instructions on how to compost your clothing, check out 1 Million Women's guide here: www.1millionwomen.com.au/blog/how-compost-fabrics/ (accessed 04/11/19).